IMAGES
of Aviation

AREA 51

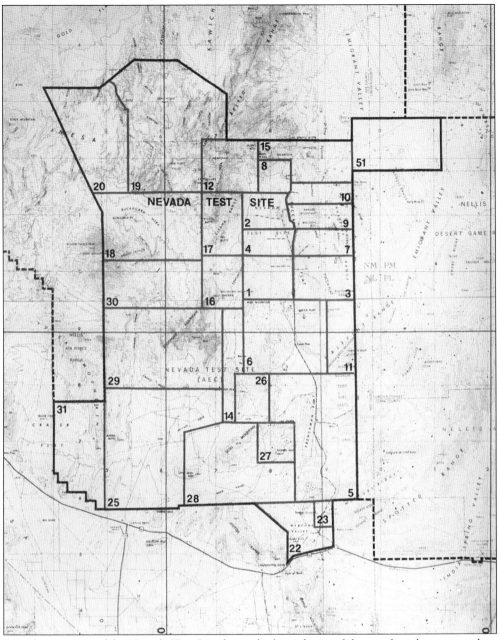

A mid-1960s map of the Nevada Test Site shows the boundaries of the numbered test areas then in use. Most of these areas were assigned numbers between 1 and 31 in no particular geographic order. A 38,400-acre block of land surrounding Groom Lake appeared on maps as early as 1957 but was not officially added to the test site until June 1958, when it was designated Area 51. (Department of Energy.)

ON THE COVER: A Lockheed A-12 sits on the runway at Area 51 awaiting departure clearance for a test or training flight while ground crewmen observe from a station wagon. In the background, a McDonnell F-101B chase plane taxis into position. At right, a second A-12 undergoes preflight checks (see page 75 for more). (Lockheed Martin Skunk Works via Jim Goodall.)

IMAGES
of Aviation

AREA 51

Peter W. Merlin

ARCADIA
PUBLISHING

Published by Arcadia Publishing
Charleston, South Carolina

Printed in the United States of America

Library of Congress Control Number: 2011926820

For all general information, please contact Arcadia Publishing:
Telephone 843-853-2070
Fax 843-853-0044
E-mail sales@arcadiapublishing.com
For customer service and orders:
Toll-Free 1-888-313-2665

Visit us on the Internet at www.arcadiapublishing.com

Technicians prepare a U-2A for flight. At right, a Military Air Transport Service (MATS) C-54G was used to bring workers to the remote Groom Lake test site. At left, the 1953 Willys-Overland one-ton, four-by-four pickup truck was one of at least six on loan from the Atomic Energy Commission. (Laughlin Heritage Foundation, Inc.)

CONTENTS

ACKNOWLEDGMENTS

When I began researching the history of Area 51 in 1983, very little information on the subject was available in the public domain. In the late 1990s, the Central Intelligence Agency (CIA) and US Air Force (USAF) started declassifying thousands of pages of documents on programs involving advanced reconnaissance platforms, stealth technology, and evaluation of foreign military hardware. Along with these documents were hundreds of photographs that provided a glimpse into a hidden world. Most important, those who once labored in obscurity were finally able to take credit for their work.

I owe a debt of gratitude to the people and agencies listed above, as well as to numerous federal and corporate historians and archivists. I especially wish to thank T.D. Barnes, president of the Area 51 alumni organization Roadrunners Internationale (RRI), for sharing his vast knowledge and photograph collection, introducing me to the unsung Cold War heroes and welcoming me as an associate member. He also spearheaded an effort to preserve the heritage of Area 51 through several websites and the Nevada Aerospace Hall of Fame, a nonprofit, educational institution dedicated to preserving the legacy of the men and women who pioneered and advanced our nation's aerospace technology within and above the state of Nevada.

I offer special thanks to the outstanding people at Arcadia Publishing, including Jerry Roberts, Debbie Seracini, Stacia Bannerman, and others without whom this project would not have been possible. Thanks, as always, to my wife, Sarah, for copyediting my draft manuscript. My gratitude also extends to fellow researchers, photographers, and historians Chris Pocock, Jim Goodall, Steve Davies, Paul Suhler, John Grayson, Chuck Clark, Michael Haywood, Peter Nicholson, Jakub Vanek, and Joerg Arnu, who is webmaster of the site Dreamland Resort (DLR) in Las Vegas (dreamlandresort.com).

Unless otherwise noted, images in this book are courtesy of the Laughlin Heritage Foundation, Inc. in Del Rio, Texas, a nonprofit organization established to inform and educate the public on the important role of air power in sustaining the national security of the United States and to preserve the heritage of Laughlin Air Force Base, including its association with the U-2. Other sources include the US Navy (USN), Department of Energy (DOE), Department of Defense (DoD), Lockheed Martin Skunk Works (LMSW), and the Stealth Fighter Association (SFA).

All material in this book is derived from unclassified sources.

INTRODUCTION

Area 51 is at once the most secret and most well-known airfield in the world. Born amid Cold War secrecy, it has become an object of mystery in modern popular culture colored by allegations of involvement with UFOs, aliens, and sinister conspiracies. Recent availability of declassified government documents and photographs makes it possible to cut through the fog of dark rumors and speculation to reveal the true legacy of one of the nation's premier assets for flight tests and evaluation of cutting-edge aviation technology.

Originally a test site for the Lockheed U-2 spy plane, the air base was built in 1955 at Groom Dry Lake, Nevada, about 84 miles northwest of Las Vegas, in sparsely populated Lincoln County. Development of the U-2 was conducted in secret because the airplane's performance characteristics and mission were classified. Although the Central Intelligence Agency was Lockheed's primary customer, the operation received a great deal of support from both the US Air Force and the Atomic Energy Commission (AEC).

Experimental aircraft and prototypes are typically tested at Edwards Air Force Base, California, on the edge of a vast dry lake bed in the Mojave Desert. For security reasons, a more remote site was required for U-2 flight operations. In an effort to find a suitable test site, Clarence L. "Kelly" Johnson of Lockheed's Advanced Development Projects division (known as the Skunk Works) sent test pilot Tony LeVier to scout potential locations. Masquerading as hunters, LeVier and Skunk Works chief foreman Dorsey Kammerer set out from Burbank, California, in an unmarked Beech Model 50 Twin Bonanza on a two-week survey mission.

Richard M. Bissell Jr., special assistant to CIA director Allen Dulles, was selected to direct the agency's U-2 program, known as Project Aquatone. Bissell reviewed 50 potential test sites with his Air Force liaison, Col. Osmond J. "Ozzie" Ritland, but found that none of them met the program's stringent security requirements.

Johnson had already picked a dry lake bed, dubbed "Site One" in his personal log, and designed an airfield to be built there. His proposal called for a small, temporary site with only the most rudimentary accommodations. As originally planned, total cost of construction would have been just over $200,000. When the CIA revised the requirements (entailing a nearly 300 percent expansion and a more permanent facility), Johnson estimated that construction costs would jump to $450,000. After Bissell rejected Site One, Ritland suggested looking at Nellis Auxiliary Field No. One, an abandoned World War II airstrip adjacent to Groom Lake on the Las Vegas Bombing and Gunnery Range.

On April 13, 1955, LeVier, Johnson, Bissell, and Ritland flew to Nevada. After illegally penetrating AEC airspace, they headed for the dry lake bed in the middle of Emigrant Valley. From the air, little evidence of human activity that had taken place in the area—including a visit by Mormon settlers seeking passage to California in 1849 and sporadic mining activity after discovery of lead, copper, and silver ore in 1864—was apparent. Ritland pointed out the X-shaped auxiliary airstrip, but LeVier chose to land on the flat surface of Groom Lake.

This proved a fortuitous decision, as the abandoned airfield was overgrown, sandy, and completely unusable—the lake bed, however, was a different story. Bissell later described the hard-packed playa as "a perfect natural landing field. . . as smooth as a billiard table without anything being done to it." It was, though, littered with shell casings and debris from Air Force gunnery practice. These had to be painstakingly removed before flight operations began.

On May 4, 1955, LeVier, Kammerer, and Johnson returned to Groom Lake and used a compass and surveying equipment to mark the location of a 5,000-foot runway just off the lake bed's southwest corner. Johnson proposed naming the base Paradise Ranch, which was an ironic choice that he later admitted was a dirty trick intended to lure workers to the U-2 project. CIA officials, however, ultimately named the test site Watertown Airstrip, but Lockheed workers continued to call it "the Ranch."

Two weeks later, Seth R. Woodruff Jr., manager of the AEC Las Vegas Field Office, announced that he had instructed the Reynolds Electrical and Engineering Company "to begin preliminary work on a small, satellite Nevada Test Site installation . . . a few miles northeast of Yucca Flat and within the Las Vegas Bombing and Gunnery Range." Woodruff said that the installation would include "a runway, dormitories, and a few other buildings for housing equipment." The facility was described as "essentially temporary." A press release, drafted for the AEC by CIA officials, was distributed to 18 media outlets in Nevada and Utah, including a dozen newspapers, four radio stations, and two television stations.

After flight-testing began in the summer of 1955, subsequent press inquiries necessitated a plausible cover story for the Watertown operation. According to several AEC press statements, "U-2 jet aircraft with special characteristics for flight at exceptionally high altitudes have been flown from the Watertown strip with logistical and technical support by the Air Weather Service of the US Air Force to make weather observations at heights that cannot be attained by most aircraft." There was no mention of the CIA or the aircraft's true mission, strategic reconnaissance.

Over the next few years, CIA and Air Force pilots underwent rigorous training at Watertown prior to deployment at locations around the world. Francis Gary Powers became the most famous member of the CIA cadre after his U-2 was shot down over the Soviet Union in May 1960. His subsequent capture and trial shed light on the clandestine aspects of the U-2 program and forever shattered the "weather research" cover story. In his autobiography, of Groom Lake Powers wrote, "As a place to live, it left much to be desired. As a secret training base for a revolutionary new plane, it was an excellent site."

Watertown personnel faced a unique challenge because the airstrip was downwind of the nuclear proving ground, making it necessary to evacuate the base prior to each detonation. The AEC tried to ensure that expected fallout from any given shot would be limited so as to permit reentry of personnel within a few weeks, but the delays were a problem. In mid-June 1957, at the beginning of an ambitious series of nuclear detonations, the last U-2 aircraft departed Watertown, and the facility was placed in caretaker status.

In June 1958, the Watertown Airstrip and surrounding land was officially joined to the nuclear proving ground (by then known as the Nevada Test Site) and designated Area 51. By this time, the CIA was already seeking a successor to the U-2, which was becoming increasingly vulnerable to surface-to-air missiles. Lockheed's Kelly Johnson again supplied a solution, designing the A-12, a high-speed, high-altitude spy plane, under Project Oxcart. Capable of Mach 3 speeds (nearly 2,200 miles per hour) and altitudes up to 90,000 feet (several times higher than those flown by commercial airliners), the A-12 was the first of Lockheed's Blackbird family of aircraft.

Again, a secret test base was needed for the new airplane, but the old airstrip was unsuitable and infrastructure for such a program was not available at Area 51. At first, the CIA considered several Air Force bases that were scheduled for closure, but none offered the required security. Area 51 was ultimately selected, although it lacked personnel accommodations, fuel storage, and an adequate runway. Lockheed project personnel submitted an estimate of requirements for monthly fuel consumption, hangars, maintenance facilities, housing, and runway specifications. The CIA then produced a plan for construction and engineering. The stage was now set to make Area 51 a permanent facility. Base construction began in earnest in September 1960 and continued on a double-shift schedule until June 1964, but the essential facilities were completed within the first year.

Air Force and CIA involvement in the project was inextricably entwined. Although the A-12 was nominally a CIA asset, Air Force Headquarters Command at Bolling Air Force Base in Washington, DC, created the 1129th Special Activities Squadron (SAS)—known as the Roadrunners—to operate the aircraft and the secret base. Project Oxcart pilots resigned military commissions for the duration of the project, becoming employees of the CIA in a process known as "sheep dipping." The aircraft, however, were painted in standard Air Force markings, including Air Force serial numbers.

Although the squadron was headquartered at Bolling, the permanent duty station for most 1129th SAS personnel was Detachment 1 in Nevada. Upon arrival in the Las Vegas area, each person called a local telephone number and received specific reporting instructions. Travel orders admonished, "Las Vegas is in excess of 100 miles from your duty station, which is a classified location." Those bringing family members were advised that Las Vegas was the closest city in which all dependent facilities and conveniences were located, and that they would normally be able to return to the city on weekends. A typical tour of duty lasted three years.

Project Oxcart spanned eight years and spawned several variant aircraft, including an interceptor capable of launching air-to-air missiles and a mother ship that launched a ramjet-powered drone for carrying out autonomous spy missions over denied territory. In early 1968—a few months before the Oxcart aircraft were retired—the Air Force began a series of projects that would have a significant impact on Area 51.

After seeing how poorly US pilots fared at the beginning of the air war in Vietnam, intelligence-community and Air Force officials developed plans to acquire examples of fighter planes used by opposition forces. A variety of aircraft types used by the enemy were evaluated at Area 51 in order to develop effective combat tactics against them. The results of these tests dramatically changed the kill ratio in favor of US forces, and these "Red Hats" programs subsequently became a high priority at the secret Nevada base.

In the late 1970s and early 1980s, the modern doctrine of "stealthy" design that effectively rendered aircraft invisible to detection by radar was pioneered at Area 51. The results of these efforts remained secret for many years but were dramatically demonstrated by the stunning allied victories during the first days of Operation Desert Storm in 1991. Several examples of Area 51's stealth prototypes now hold places of honor at the National Museum of the US Air Force in Dayton, Ohio.

In 1977, the CIA transferred responsibility for the Groom Lake base to the Air Force. Initially, a small cadre of personnel from the Air Force Flight Test Center at Edwards Air Force Base supervised the Red Hats and stealth projects. Two years later, the organization was formally designated Detachment 3, AFFTC. By the 1990s, the detachment had grown into a wing-sized national test facility with an annual operating budget in excess of $200 million and responsibility for conducting numerous test programs, including high-priority, presidentially directed classified efforts.

Surrounded by mystery, Area 51 has become an icon of popular culture. The secret base has been featured in books and magazine articles, video games, Internet sites, and numerous films and television documentaries, including blockbuster movies like *Independence Day* and shows such as *The X-Files*. Operational details of many projects that have taken place at Area 51 remain classified due to national security. Most of the material now publicly available on particular programs was only declassified many years after they had been concluded and were no longer deemed sensitive. This book provides a rare glimpse into what is sometimes evocatively referred to as the "black world."

Throughout the 1950s, Watertown Airstrip was a spartan facility comprised of a single 5,000-foot asphalt runway on the southwest corner of Groom Lake, hangars, and other buildings. The dry lake bed also served as a landing field. An initial cadre of 75 test personnel grew to 250 during U-2 training operations. The base population peaked at 1,250 persons. (USGS.)

One

WATERTOWN

HOME OF THE DRAGON LADY

By July 1955, construction at Watertown was progressing quickly, but Kelly Johnson feared the base would not be ready in time to meet the first flight scheduled for a U-2. He authorized the use of Lockheed funds to procure a trailer for the test team and arranged for the acquisition of a C-47 transport and two T-33 jet trainers. A drilling crew working near the edge of Groom Lake discovered a limited source of water but encountered difficulty establishing a functioning well. Despite these troubles, Johnson felt that the government was getting a bargain for its money.

The first U-2 airframe arrived as disassembled on July 25 in a C-124 cargo plane. Test flights commenced at the end of the month, and soon more Project Aquatone aircraft were on their way to be readied for use in training CIA pilots. Meanwhile, the Air Force ordered 29 U-2 aircraft that would be flown by the Strategic Air Command under project Dragon Lady.

In September 1955, John M. Ferry, special assistant for installations at the Department of the Air Force, informed the director of the Bureau of Land Management that the AEC had "an urgent requirement for the extension of the Nevada Test Site, which is used in connection with continuing experimental sampling projects." Ferry noted that the Air Force had determined that the 60-square-mile tract of land surrounding Groom Lake was no longer required for use as a bombing and gunnery range, and it was returned to the Department of the Interior. Subsequently, AEC officials submitted a formal request to add the parcel to the Nevada Test Site.

By early March 1956, the first CIA pilots were undergoing flight training; classes for the first Air Force U-2 pilots began nine moths later. Meanwhile, Lockheed pilots continued to test new systems and capabilities for the aircraft. The CIA deployed its U-2 fleet to various detachments around the globe. In June 1957, test operations moved to Edwards while the Air Force redeployed its 4028th Strategic Reconnaissance Squadron to Laughlin Air Force Base at Del Rio, Texas.

Each U-2 arrived as disassembled and inside a C-124 transport. With a maximum takeoff weight of 216,000 pounds, landing usually took place on the lake bed to avoid wear and tear to the paved runway. When the first U-2 was delivered on July 25, 1955, base commander Col. Richard Newton initially refused to allow the C-124 to land on the asphalt strip after rain had softened the lake-bed surface. He relented after Kelly Johnson candidly expressed his dissatisfaction and called CIA Headquarters. Two hours later, using reverse propeller pitch for braking, the C-124 landed on partially deflated tires. After the dust cleared, the commander noted that the runway was indented to a depth of a quarter inch for a distance of 50 feet. "It was real gory for a first meeting with Newton," Johnson later wrote in his personal log.

Air Force crews unload a disassembled U-2 from a C-124 parked on the lake bed. Tony LeVier and fellow Lockheed test pilot Bob Matye spent several weeks removing surface debris from the playa to make it usable as an airfield. LeVier also drew up a proposal for four three-mile-long runways to be marked on the hard-packed clay, but Kelly Johnson at first refused to approve the $450 expense, citing a lack of funds. Several lake-bed runways were eventually added, allowing flight operations regardless of wind direction.

The U-2 prototype had no military serial number and was simply designated Article 341. Early in the test program, it bore only the US national insignia and tail no. 001. Built entirely of aluminum alloys, the U-2 was extremely lightweight and was equipped with an unusual bicycle landing-gear configuration. (LMSW.)

Lockheed test pilot Tony LeVier made the unofficial maiden flight during a taxi test on July 29, 1955, leaving the ground at a speed of 70 knots and remaining airborne for a quarter-mile before making a rough landing. He was tasked with completing initial contractor testing, taking the U-2 to 50,000 feet, achieving the maximum design speed of Mach 0.84 and demonstrating a successful "dead-stick" landing, where the aircraft loses its power and is forced to land. (LMSW.)

LeVier made the first intentional flight on August 4, 1955, taking off from the north end of the lake bed in line with the Watertown runway. Article 341 climbed to an altitude of nearly 8,000 feet through a light rainstorm just north of the airfield. The airplane flew beautifully, but landing proved difficult. The pilot made several attempts before finally touching down just ahead of a downpour that flooded the lake bed with two inches of water. LeVier left the U-2 program after completing 19 additional test flights. Bob Matye and Ray Goudey expanded the airplane's altitude envelope to 74,500 feet. By November 1955, the test group also included Robert Sieker and Robert Schumacher. (LMSW.)

With few test aircraft available, the Watertown flight line, seen here in 1955, was vacant except for a fleet of utility vehicles and a fire truck. On November 2, Maj. Gen. Albert Boyd from Air Research and Development Command and Lt. Col. Frank K. "Pete" Everest Jr. from the Air Force Flight Test Center arrived for orientation flights in the U-2. Everest reportedly had difficulty landing the airplane. (Bob Murphy.)

Watertown's three "T" hangars are flanked by the control tower and a warehouse. By March 1956, nine U-2 aircraft had been delivered to the test site. Col. Landon McConnell replaced Newton as base commander, and Col. William Yancey oversaw flight-training activities. Four instructors trained pilots in ground-school classes, which were followed by landing practice in a T-33 and, eventually, solo flights in the U-2.

In order to conceal the true purpose of the U-2, Hugh L. Dryden, director of the National Advisory Committee for Aeronautics, issued a press release in May 1956 announcing a program in which the aircraft would conduct high-altitude weather research for the NACA with Air Force support while operating from Watertown Strip. To support this cover story, the airplanes were painted in NACA markings.

Working near the edge of the lake bed, technicians prepare a U-2 for flight. During test and training flights, the aircraft carried NACA weather research instrumentation to collect data on turbulence above 50,000 feet and other high-altitude phenomena. The NACA published the results in several unclassified reports.

Watertown Airstrip, viewed from the Papoose Mountains in April 1957, accommodated Lockheed's Beech Twin Bonanza, nine U-2 aircraft, four T-33A trainers, and a C-47 transport. An L-20A liaison aircraft is just visible on the left. Base facilities included a 5,000-foot runway, control tower, three hangars, a few warehouses, administrative buildings, dining hall, about three-dozen trailers, and a water tower. The base's few amenities included a movie theater and volleyball court.

Although construction costs totaled about $832,000, more than twice the amount Kelly Johnson had estimated for a similar facility at his original location choice, he felt the government had gotten a very good deal. Watertown's proximity to the atomic proving ground meant the base lay directly in the downwind path of radioactive fallout from aboveground nuclear tests, but AEC perimeter and airspace restrictions helped shield the operation from public view.

Several dozen trailers provided rudimentary accommodations for Watertown personnel. This scene illustrates the most mundane details of life on a secret installation. After leaning his push broom against the side of a trailer, a custodian carrying a wastebasket stoops beside a pile of bed linens destined for the base laundry.

Security guard Richard Mingus climbs approximately 75 steps to reach the top of the base water tower, which doubled as a guard post. It provided a commanding perch from which to observe activities on the flight line and in the housing area. The tower also served as a convenient place to mount various antennas.

Federal Services Inc. (FSI) provided security for the Nevada Test Site and Watertown. Guards were tasked with physical protection of the test site and its structures, equipment, materials. They were also charged with maintenance of access controls over all persons entering the base, ensuring that all personnel admitted possessed the required security clearance and limiting access only to those areas where their presence was required.

The elevated guard post provided a birds-eye view of the spartan accommodations on the south side of the base. Most personnel occupied 30-foot-long mobile homes, each housing three people. In addition, there was a dormitory with 18 bedrooms adjacent to the headquarters building.

Watertown personnel unload luggage, including a pair of cross-country skis, from a MATS C-54G transport. Louis Setter, who served as a U-2 instructor pilot, recalls attempting to tow a ski enthusiast behind a single-engine L-20A airplane while taxiing across the snow-covered lake bed. The results were unsatisfactory due to blowing snow from the prop wash.

After disembarking from the commuter shuttle, passengers line up for a security check. The Douglas C-54 Skymaster could carry as many as 60 passengers. The MATS flights were temporarily suspended after one of the aircraft crashed on Mt. Charleston en route to Watertown from Burbank in November 1955, resulting in the loss of nine civilians and five military personnel.

Working from the back of a four-by-four pickup truck, a steely-eyed FSI security officer diligently checks the identities of each person entering the base. Every worker had to check in upon arrival and pick up a special access badge that was turned in upon departure. This procedure ensured that only authorized personnel gained access to restricted areas and that the identity of all visitors was duly recorded.

Ground-crew personnel assist a pilot preparing for a flight in a U-2A. These photographs, taken through the cockpit and passenger windows of a C-54 transport, highlight the photographer's artistic flair. The flat expanse of Groom Lake is visible in the background, along with the Groom Mountains and the 9,380-foot peak of Bald Mountain.

A technician works on a Type B camera with a 36-inch focal length. Used for large-scale, high-resolution photography, the Type B-1 could resolve features as small as 2.5 feet across from an altitude of 65,000 feet. The lens cone pivoted to allow horizon-to-horizon exposures. (LMSW.)

The Type A-2 camera package had to be winched up into the U-2's payload bay, or Q-bay, which spanned the entire fuselage just aft of the cockpit. This package, consisting of three 24-inch focal-length cameras on fixed mounts, was used on early operational reconnaissance missions. Each camera carried 1,800 feet of film and was capable of resolving objects two to eight feet across. (LMSW.)

Article 341's lower fuselage was covered with radar-absorbent material during Project Rainbow. A fiberglass honeycomb sandwich, varying in thickness from a quarter inch to about one inch, was topped with layers of Salisbury Screen. The canvas was painted with a conductive graphite grid. It was nicknamed "wallpaper" because of the circuit pattern and "thermos" because it acted as insulation that prevented dissipation of engine heat through the aircraft's skin. (LMSW.)

During a Project Rainbow test flight on April 4, 1957, airframe heat build-up caused engine flameout at 72,000 feet. Lockheed test pilot Robert Sieker's pressure suit inflated properly when cabin pressure was lost, but the clasp on his faceplate failed, resulting in a loss of oxygen that caused him to temporarily lose consciousness. He died while attempting to bail out at low altitude. Searchers located the crash site three days later. (CIA.)

Another attempt to make the U-2 less visible to radar involved stringing copper-plated steel wires with ferrite beads across the aircraft's outer skin at specific distances from the fuselage, wings, and tail. Nicknamed "trapeze" and "wires," this treatment was applied to Article 343. (LMSW.)

With the addition of radar-absorbent structures and coatings, the U-2 suffered from excess weight and drag, making it aerodynamically "unclean." Hence, the Project Rainbow test beds were known as "dirty birds." Although these treatments were effective against some radar frequencies, they reduced the airplane's operating ceiling by 1,500 to 5,000 feet and cut its range by 20 percent. (LMSW.)

The flag in front of Base Headquarters was lowered to half-staff following the death of Robert Sieker. Building 104 encompassed approximately 3,800 square feet of space devoted to administrative offices, conference rooms, equipment storage, and showers for flight crews. Because the airstrip was downwind of the nuclear proving ground, radiation safety monitors from the Nevada Test Site documented the materials and structures in this and other buildings at Watertown and installed instrumentation to measure the effects of radioactive fallout. In effect, Watertown served as a laboratory to determine the shielding qualities of building materials that might typically be found in any American small town. Watertown personnel were required to evacuate the base prior to each detonation, and the AEC tried to ensure that expected fallout from any given shot would be limited so as to permit reentry of personnel within three to four weeks. Evacuation plans included provisions for securing classified areas, a system for informing evacuees as to when they might return, and radiation monitoring. All personnel were required to wear radiation badges to measure their exposure to fallout.

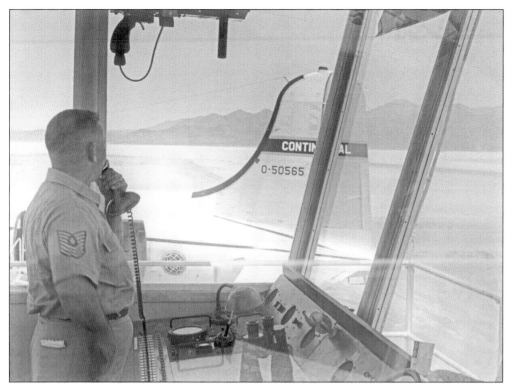

Air Force personnel controlled the movement of aircraft on and around Watertown Airstrip. The control tower, located just north of Base Headquarters, was equipped with UHF and VHF radios and equipment to measure wind speed and direction. In the event of loss of communication due to radio failure, a light gun, seen here hung from the ceiling, was provided for signaling incoming aircraft. Flight operations took place using the paved airstrip as well as runways marked on the clay surface of the dry lake bed. Commuter flights from Burbank and Las Vegas were vectored to Camp Desert Rock near Mercury before turning north toward Groom Lake.

The O-11A fire truck, built by American LaFrance, carried 1,110 gallons of a water-foam solution for extinguishing blazes fed by gasoline or jet fuel and 40 gallons of chlorobromomethane, a vaporizing liquid used to extinguish fires in engine nacelles and other confined areas. The vehicle was equipped with two remote-controlled turret nozzles above the cab, each capable of discharging 200 gallons per minute, and had provisions for hand lines.

For response to medical emergencies, the Air Force provided a 1954 Henney-Packard Junior ambulance. This model was commercially unpopular, and the company ended up selling a substantial number to the government at a loss. A three-quarter-ton, four-by-four 1953 Dodge M-43 military ambulance is visible in the background.

Maintenance technicians work into the night to assemble a U-2 in one of the three Watertown hangars. They appear to be working on the attachment joint for the left wing. A 1952 Willys M38 Jeep is parked in front of the building on the left.

Ford Country Sedan station wagons (a 1957 model [left] and a 1955 model) served as Mobiles, chase cars that followed the U-2 during takeoff and landing. An experienced pilot with a radio rode in one of the cars to advise the U-2 pilot with critical information such as altitude above the runway.

The AEC loaned this de Havilland L-20A Beaver to Watertown for use as a spotter plane to patrol the test site for intruders. One day, as Louis Setter landed the plane on the snow-covered lake bed, he saw a man on skis. It was a worker from the U-2 project who requested that Setter try towing him behind the plane while taxiing. (DOE.)

Lockheed's Beech Model 50 Twin Bonanza served a multitude of purposes. In early 1955, it was flown during a survey of potential test sites. Lockheed officials used it for transportation between Burbank and Watertown throughout the program, and it served as a chase plane during test flights.

The U-2 was designed for ease of assembly and disassembly. The tail section was joined to the fuselage with only three 5/8-inch bolts. A single J57 turbojet engine was located in the center of the fuselage in line with the wings. The exhaust was ducted through an insulated tailpipe. Here, a technician is barely visible as he leans into the mid-fuselage to gain access to the engine.

The flight surgeon, seen here in the Watertown dispensary, played a vital part in preparing U-2 pilots for their missions. The health and fitness of each pilot was carefully monitored. The flight surgeon was responsible for performing a medical checkup before each flight in addition to supervising oxygen pre-breathing and other physiological support activities.

Ground crewmen prepare Article 364 for a training mission. The U-2 was a difficult aircraft to fly, and two CIA pilots lost their lives during training in 1956. The first fatality occurred on May 15, when Wilburn S. "Billy" Rose tried to shake a hung pogo loose after takeoff, stalled, and crashed. Frank G. Grace Jr. was killed during an August 31 night training flight when he became disoriented by lights near the end of the runway and flew into a telephone pole. On December 19, Robert J. Ericson suffered an oxygen failure at 35,000 feet over Arizona but parachuted to safety.

With the Groom Range and Bald Mountain in the background, Article 361 sits on the lake bed after landing. Ground crewmen work on the aircraft while the pilot sits in the Mobile. A 1953 Willys-Overland one-ton, four-by-four truck waits to tow the aircraft back to Watertown Airstrip. Article 361 was the first U-2 built for the Air Force and was constructed at Lockheed's plant in Oildale, California.

Early test and training flights were restricted to within 200 miles of Watertown, well within the glide range of a U-2 from high altitude in the event of engine failure. After gaining more confidence in the aircraft, planners authorized pilots to fly farther during long-duration flights that often lasted eight hours or more.

Article 365 sits on a transport dolly during an engine run. The U-2 was equipped with a Pratt & Whitney J57-P-31 turbojet engine that produced 11,200 pounds of thrust. At altitudes above 70,000 feet, it burned just 700 pounds of fuel per hour. It was not prone to leak oil or flame out but could not be restarted above 37,000 feet.

A maintenance crew works on Article 368; one man is on a ladder in the Q-bay and another is working on one of the nose antenna panels. Behind the gray fiberglass panels, most early U-2s carried equipment for collecting and recording electronic and communications signals on several frequencies.

At 9:48 a.m. on April 9, 1957, a pilot fills out paperwork in the Watertown operations office. According to the schedule board, a C-54G is inbound from Burbank. Bennedict Lacombe (call sign Workhorse 17) had landed 30 minutes earlier following an eight-hour flight in Article 361. Raymond Haupt (Workhorse 16) took off at 8:04 a.m. for a training sortie in Article 365, followed by Donald Sorlie in Article 368.

T.Sgt. Weldon C. Lewis checks Maj. Richard Heyser's inner flight helmet assembly. Also visible are helmets belonging to Col. Jack Nole and Maj. Joe Jackson. The model MA-2 helmet, made by the International Latex Corporation of Dover, Delaware, provided constant pressure and oxygen to the head with a tight-fitting liner and seal covered by a hard shell.

Technicians work on a "sniffer" package for installation in Article 372. The U-2 was capable of carrying a variety of particulate sensors, including a sampling system consisting of six 16-inch filter papers that were exposed to air fed from a duct faired onto the left side of the Q-bay hatch.

Article 373 awaits final assembly shortly after delivery to Watertown. Darkrooms and camera maintenance shops occupied the south end of each T-hangar. Several desks and file cabinets are visible on the upper floor, along with a map of the flight line indicating aircraft movements and a couple of nude pin-ups.

The partially disassembled airframe of Article 373 rests on transport dollies inside one of Watertown's three T-hangars. The U-2 was easy to assemble and disassemble for transport due to its modular construction and the relatively small number of fasteners required.

The U-2 could be quickly assembled by a handful of workers. Here, one team of technicians prepares to install the vertical tail atop the empennage, while a second group has just finished mating the engine to its exhaust duct.

For transport to the test site, the U-2 could be broken down into several manageable subassemblies, including wings, tail surfaces, fuselage, empennage, and propulsion system. Extra care had to be taken to avoid damaging the airplane's thin aluminum skin.

Maintenance technicians prepare a U-2 fuselage assembly for engine installation; the wings have yet to be attached. Each airplane was assembled shortly after arrival at Watertown and subjected to functional testing followed by acceptance test flights.

At 1:05 p.m., technicians pause in their tasks. The wings of this U-2 have been mated to the fuselage, but several panels still need to be attached. Pads cover portions of the aluminum skin to prevent damage from accidental tool strikes.

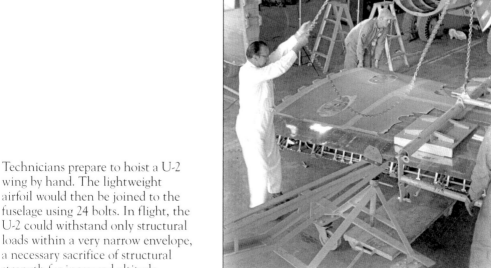

Technicians prepare to hoist a U-2 wing by hand. The lightweight airfoil would then be joined to the fuselage using 24 bolts. In flight, the U-2 could withstand only structural loads within a very narrow envelope, a necessary sacrifice of structural strength for increased altitude.

Flight planning was a team effort and the key to success for U-2 missions. In order to maximize target coverage, planners at Watertown carefully studied meteorological data and topographic maps, selected a flight route, and chose appropriate cameras and sensors. In these photographs, taken on April 9, 1957, men with expressions of intense concentration pore over mission details. Maps and other documents are spread across furniture borrowed from the AEC. In a back room in the image below, classified materials are stored in filing cabinets with combination locks. The door between the two rooms hangs open despite a sign reading, "Briefing in progress. Do not enter."

Article 372 undergoes maintenance in the hangar with Article 368 just outside. Work was occasionally interrupted by atomic tests taking place just over the hills at Yucca Flat. All personnel at the base were required to wear radiation badges to measure exposure to fallout, and Watertown personnel were evacuated prior to each scheduled shot date. On May 14, 1957, AEC radiological safety officers briefed Watertown personnel on nuclear testing activities, radiation safety, and the possibility of radiation hazards from the upcoming Operation Plumbbob test series. The film *Atomic Tests in Nevada* was shown and discussed.

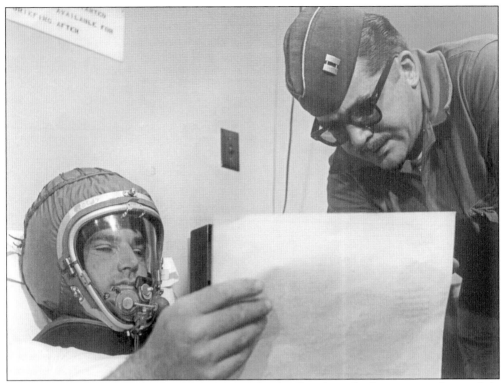

Lockheed test pilot Ray Goudey receives a weather briefing from the base meteorologist while pre-breathing oxygen in preparation for a high-altitude flight. Before each flight, U-2 pilots were required to breathe oxygen for two hours in order to eliminate nitrogen bubbles from the bloodstream; otherwise, they could experience a painful phenomenon similar to the "bends" that is sometimes suffered by deep-sea divers during decompression. Pilots typically used their pre-breathing time to catch up on mission briefing material or simply relax. Weather briefings were particularly important, as the lightweight structure of the U-2 was not designed to withstand heavy turbulence.

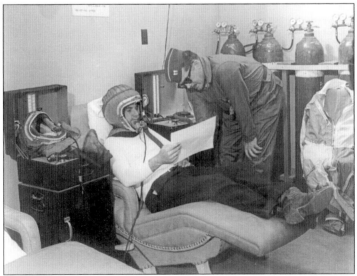

While pre-breathing oxygen in the Watertown life-support facility, Ray Goudey reads the 1955 edition of *Adventures in Time and Space*, a collection of short stories by noted science-fiction authors of the day. Here, in the heart of a secret airbase that would one day be a focal point for fantastic rumors of aliens and flying saucers, Goudey enjoyed such contemporary science-fiction tales as "The Proud Robot" by Lewis Padgett, "Heavy Planet" by Lee Gregor and Frederik Pohl, "The Link" by Cleve Cartmill, "Adam and No Eve" by Alfred Bester, "Nightfall" by Isaac Asimov, "The Roads Must Roll" by Robert A. Heinlein, and "Within the Pyramid" by R. DeWitt Miller. Vertical lines on Goudey's faceplate are heating elements that defrost the glass.

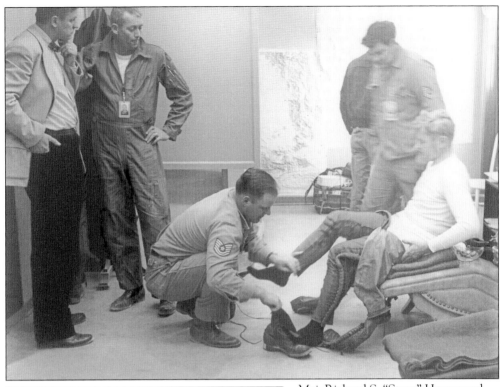

Maj. Richard S. "Steve" Heyser and an unidentified civilian look on as life-support technicians assist a U-2 pilot in removing his flight gear. Physiological-support equipment worn by the pilot creates an environment designed to minimize the impact (physical and physiological) of flying at extreme altitudes. A partial pressure suit was necessary above 50,000 feet for protection from such hazards as hypoxia and decompression sickness.

Early U-2 pilots wore MC-3 partial pressure suits developed by the David Clark Company of Worcester, Mass. The MC-3 (usually worn beneath a light coverall) was the first suit to employ the capstan principle, using inflatable tubes and cross-stitching to automatically tighten the suit against the pilot's body if the cockpit depressurized. This produced mechanical pressure to counteract the expansion of gases and fluids in the body at high altitude.

A life-support technician checks oxygen equipment while Lockheed test pilot Robert Schumacher undergoes oxygen pre-breathing prior to flight. Schumacher was the fifth test pilot to join the program, arriving at Watertown in mid-November 1955. He eventually went on to test an advanced model called the U-2C as well as the first two-seat U-2, known as the U-2D.

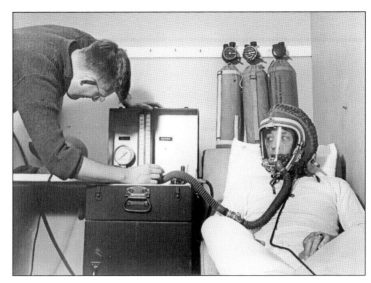

Maj. James A. Qualls pats Col. Jack Nole, commander of the 4028th Strategic Reconnaissance Squadron, on the back as ground crewmen assist him with last-minute adjustments to his pressure suit and outer coveralls. The cylinder provided breathing oxygen before the pilot was hooked up to the airplane's life-support systems.

Relaxed postures belie the serious nature of a preflight briefing prior to a long-duration U-2 mission. In the image above, Maj. Richard Heyser stands at the podium. A navigation chart of the United States, with all Air Force bases highlighted, serves as a backdrop. A sign on the wall at right reads, "Reason? Hell, there ain't none. It's just company policy." Below, a meteorologist presents the latest weather forecast predicting light turbulence and contrail formation between 27,000 and 37,000 feet and a cold front pushing down through the central states. A photographic map (far left) shows the layout of Watertown Airstrip and several lake-bed runways.

Above, an Air Force meteorologist checks instruments in the Watertown weather office, conveniently located near the flight line. Below, a pilot studies the latest forecast for weather conditions and winds aloft. Groom Lake offered excellent flying weather almost year-round, but U-2 training flights often spanned the country, exposing the airplane to a wide variety of meteorological conditions, such as air turbulence, that posed a serious danger to the airplane. Contrails at high altitude revealed the position of the U-2 to observers on the ground, which could be a problem if the airplane was flying over hostile territory. A poster on the wall warns, "Compare face with badge. Secure your area!"

An aerial photograph taken on August 28, 1968, shows how much the base had expanded. The original 5,000-foot-long airstrip had become a taxiway. The new runway consisted of an 8,625-foot-long concrete strip plus a 6,000-foot-long asphalt extension to a concrete turnaround pad, followed by another 5,000 feet of asphalt overrun. The asphalt sections were not lighted for nighttime use. A semicircle (called the "Hook") approximately two miles in diameter was marked on the dry lake so that an A-12 pilot approaching the end of the overrun could abort to the hard-packed playa if necessary rather than run his aircraft into the sagebrush. Two unpaved runways marked on the lake bed provided alternatives during heavy crosswinds. The radar cross-section measurement range is visible along the west side and southwest corner of the lake bed. On the north parking ramp, a B-52H has just returned from launching a D-21B drone over the Pacific Missile Range. Two commuter transports, an L-1049 Super Constellation and F-27F, are parked on the "Southend" apron. (USGS.)

Two

AREA 51

THE BLACKBIRD'S NEST

Watertown Airstrip was officially added to the Nevada Test Site as Area 51 in June 1958. Seventeen months later, the AEC announced construction of new facilities for "Project 51" at Groom Lake for the purpose of housing data-reduction equipment for use by Edgerton, Germeshausen and Grier, Inc. (EG&G). A subsequent article in the *Las Vegas Review Journal* noted that the Groom Lake area was "ideally suited to secret projects because experimental aircraft can take off and land without detection from any outside point."

This seemed to make it a perfect test site for Project Oxcart, but not everyone was satisfied. Following a three-day visit in 1961, CIA Inspector General Lyman B. Kirkpatrick was highly critical of Area 51 security, remarking, "The 'Area' in my opinion appears to be extremely vulnerable, in its present security provisions, against unauthorized observation. The high and rugged northeast perimeter of the immediate operating area, which I visited in order to see for myself, is not under government ownership. It is subject to a score or more of mineral claims, at least one of which is visited periodically by its owner. Several claims are sites of unoccupied buildings or cellars, which, together with the terrain in general, afford excellent opportunity for successful penetration by a skilled and determined opposition." Surprisingly, this situation was not remedied for more than two decades.

Test flights of the A-12 began in April 1962 and continued through June 1968. An interceptor variant, the YF-12A, was flown at Area 51 before being moved to Edwards Air Force Base in 1964. Additionally, the M-21 mother ship and D-21 drone underwent several test flights before a fatal accident necessitated modifying the drone for launch from a B-52. In 1968, the Air Force and Navy began the first of a series of projects to test and evaluate foreign military aircraft, radar, and weapons. The 6512th Test Squadron's Special Projects Branch at Edwards initially managed these efforts. Due to its diversified and expanded mission, classified nature, and geographic separation from its parent unit, the Special Projects Branch was elevated to full squadron status in December 1977 as the 6513th Test Squadron, dubbed "Red Hats."

James A. Cunningham Jr., director of the CIA's Development Projects Division, described the airstrip as "a monstrous runway that went all the way from hell to breakfast." The four large hangars on the North Ramp housed two A-12 test aircraft, the two-place trainer, and other projects. The operational Oxcart aircraft occupied the Southend hangars. Other facilities included workshops, offices, warehouses, a commissary, and fire stations. A three-hole golf course is visible in the lower left.

Workers lived in mobile trailers and permanent housing units called "hooches." The new Headquarters building was located near the center of the complex. Recreational facilities, seen here to the right, included tennis courts and an indoor swimming pool. The dark spot in the upper right is the "Crash Pit," used for training firefighters. (RRI.)

This view, taken around 1964, shows the main cantonment area looking from the east. Kelly Johnson's L-329 JetStar is parked just east on the large hangars on the North Ramp that housed A-12 aircraft allocated for testing, as well as the YF-12A and M/D-21. An L-1049 Super Constellation is parked just to the north.

This image, also taken around 1964 and looking south, depicts the EG&G radar building on the edge of the lake bed. The Southend hangars that housed operational A-12s are visible in the distance. Aircraft parked on the South Ramp include F-101 Voodoo chase planes and a C-47. By this time, the original Watertown T-hangars have been converted into photographic and instrumentation labs, machine shops, and maintenance areas for ground equipment. (RRI.)

Support aircraft at Area 51 included a U-3B (left) and a Cessna 210D. In 1964, Lt. Col. Burt Barrett, 1129th SAS training and operations officer, asked Charlie Trapp to pick up the Cessna after it was delivered to Las Vegas by the manufacturer. Trapp checked himself out in the plane on the way back to Area 51. (RRI.)

Standing beside a 1961 Ford F-250 four-by-four pickup, the Area 51 fire chief observes a hazardous aircraft fueling operation. The base fire department consisted of 21 men working in two shifts—four days on duty and four days off. Firefighters were always on standby during tests involving aircraft fuel tanks, engines, or the twin-Buick V-8 engine starter units used for the A-12, as well as during takeoffs and landings. (LMSW.)

The hooches, officially known as Babbitt Housing, provided spartan accommodations only marginally better than the trailers. To occupy non-duty hours, Oxcart project pilots converted House 6 into a bar and established a running poker game. Other recreation facilities included a three-hole golf course established by base commander Col. Hugh "Slip" Slater, a movie theater, gymnasium, basketball and squash courts, a fishing pond called "Slater Lake," and a softball field. With players like Harry Martin, Tom Lewis, and "Big Clem" Byzewski, the Area 51 Eight-Ballers routinely crushed such opposing teams as the Nevada Test Site's Area 12 Mets and Mercury All Stars. (RRI.)

Few minorities worked at Groom Lake in the 1950s and 1960s, but the most fondly remembered of those who did is Murphy Green of REECo Culinary Services. He reportedly ran the mess hall with an iron fist, often scolding officers and contractors—regardless of rank—for violating one of his rules. He also provided an excellent meal selection, including steak and seafood, which was much appreciated at the remote outpost. He poses here as a fisherman for a gag photograph of the flooded west side of Groom Lake on April 6, 1965. Below, Green dispenses champagne into paper cups for an October 3, 1966, celebration of the 500th A-12 flight. (RRI.)

Detachment 1, 3rd Weather Wing, commanded by Maj. Ralph W. "Bill" Thomas (bottom row, second from left), provided meteorological data to flight planners at Area 51. Thomas served with the unit from 1965 to 1968, deploying to Kadena Airbase, Okinawa, Japan, for Operation Black Shield. This photograph was taken at Area 51 on July 2, 1968. (RRI.)

The Area 51 Security Force wore this type of badge during the 1960s and early 1970s. Security personnel were responsible for protecting facilities, patrolling the base perimeter, controlling access to restricted areas, responding to actual or suspected sabotage, implementing radiological defense plans, and ensuring that sensitive assets remained under cover when foreign reconnaissance satellites passed overhead. (Author's collection.)

Following two years of test flights, the CL-329 JetStar prototype was reassigned to serve as Lockheed's Burbank-based corporate jet. Kelly Johnson used it for trips to Area 51, often accompanied by VIP guests. Unlike the four-engine production model, the prototype CL-329 was powered by two 4,850-pound-thrust Bristol Orpheus engines, of which only four were ever built. (LMSW.)

Lockheed's fleet included three Constellation-type aircraft, each capable of carrying 99 passengers. In 1960, Lockheed sold the prototype L-1649 Starliner to the CIA for $305,000, including $37,000 to prepare it for transporting personnel and cargo to Area 51. In 1963, two L-1049 Super Constellations joined the shuttle fleet. During the Oxcart program, the three aircraft made 11,495 flights, carrying 492,205 passengers and 4,328,073 pounds of cargo. (LMSW.)

Ejection-seat testing for the A-12 consisted of towing a mock-up of the nose section and cockpit behind a rented 1961 Ford Thunderbird speeding across Groom Lake. As the seat, occupied by an anthropomorphic dummy, blasted into the air, it was filmed with a motion-picture camera attached to the mock-up. (LMSW.)

For water-survival training, Oxcart pilot Ken Collins practices a parachute landing in base's indoor swimming pool. Life-support technicians helped Collins don his pressure suit, hoisted him with a special sling, and then dumped him into the water. In the background, helicopter pilot Capt. Charles E. Trapp Jr. is talking to one of the rescue personnel. (RRI.)

In September 1959, EG&G agreed to move its radar test facility from Indian Springs to Area 51 in order to perform radar cross-section testing of A-12 scale models and full-size mock-ups. On November 17, an AEC spokesman announced, "Sheet metal workers needed at the Groom Lake Project 51 in the Nevada Test Site are constructing a Butler-type building" to house data-reduction equipment for use in an Air Force program. The announcement was made because of publicity generated by a labor dispute. The sheet-metal workers union complained that the contract had been negotiated without being let for bid. REECo, the primary contractor for the AEC, obtained a court order to force the union to provide six sheet-metal workers for the top-secret project before agreeing to arbitration of the dispute prior to an injunction hearing in district court. (RRI.)

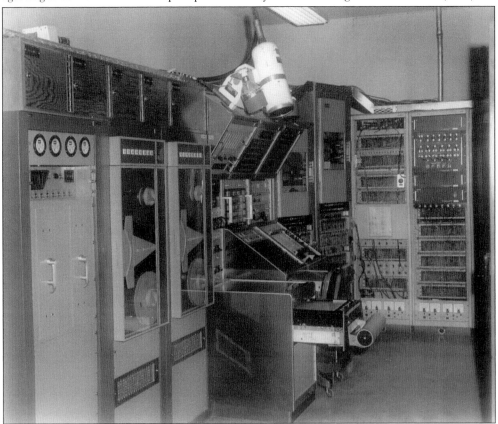

Data from the 60-foot-tall dish antenna and smaller radar arrays were fed into various equipment supplied and installed by EG&G Special Projects in Las Vegas. The company was responsible for operating the radar, performing maintenance, and data reduction. To improve efficiency and keep work on schedule, dual sets of recording equipment allowed for simultaneous testing at two frequencies. Tools of the trade included amplifier and recorder systems, charts and logs for displaying data, and timers to warm the equipment up prior to tests. (RRI.)

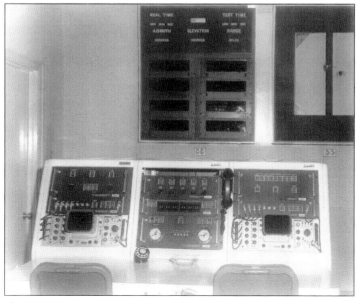

Lockheed built a 1/8-scale model of the A-12 for RCS testing and mounted it on a 22-foot-long inflatable bag within line of sight to the EG&G radar at a distance of half a mile. The bag was nearly transparent to radar, making it preferable to a shielded metal pole but incapable of supporting the weight of a full-size model. Straight seams on the bag caused measurable radar reflections. (LMSW.)

The inflatable pylon sat in the middle of a pit nicknamed "the swimming pool." The straight-seamed bag was eventually replaced by one with a spiral seam. The test model included structures to simulate afterburner exhaust plumes that presented a substantial radar return from the airplane's aft quadrant. (LMSW.)

Several full-scale A-12 models of varying configurations were tested on a 50-foot-high pole in the center of a concrete pad on the west side of the lake bed, one mile north of the EG&G radar building. The pole, designed by Lockheed engineer Henry Combs, was built from three propeller shafts of the type used on Navy destroyers, welded end to end. A piston-actuated ram elevated the pole, raising the model into the air atop a rotating head designed by Lockheed's Leon Gavette. The framework lying on the ground in the photograph below was part of a marginally successful attempt to shield the pole to prevent radar backscatter. (LMSW.)

Technicians mate the full-scale model to the pole and rotator. This early version of the A-12 had swept-back vertical stabilizers that were not incorporated into the final design. In the background, the base looks much as it did during the U-2 program. The large hangars on the North Ramp had not yet been constructed. (LMSW.)

A man standing next to a Willys M38 Jeep provides scale in this photograph of a full-scale A-12 mock-up in its final external configuration, with all-moving rudders on stub fins. For some reason, the model has been entirely painted in a camouflage pattern, perhaps to conceal its shape from ground observers. (LMSW.)

RCS testing went on day and night in order to meet an ambitious schedule after the CIA approved 48-hour workweeks for EG&G personnel. Because the A-12's exhaust ejectors acted like corner reflectors and produced a large radar return, Lockheed physicist Ed Lovick suggested ionizing the exhaust gases in order to shield the ejectors from most of the incoming radar energy. Working with engineers from Pratt & Whitney, he experimented with potassium, sodium, and cesium fuel additives to find a metallic salt that, when vaporized, had very low ionization potential and efficiently produced a plasma cloud. To simulate the results during RCS testing, the pole model was fitted with cylindrical frameworks covered with reflective material that mimicked radar returns from the ionized exhaust plumes. A special shield prevented backscatter from the pole and rotator. (LMSW.)

The full-scale A-12 model with radar-absorbent structures in the chines, wing edges, tails, and inlet spikes was mounted on the 50-foot-tall pole in November 1960. The airplane's anti-radar treatments consisted of graphite-loaded asbestos-silicone honeycomb sandwiched between laminated asbestos sheets. The configuration was optimized to defeat frequencies associated with the Soviet P-14 (NATO code name "Tall King") long-range tracking radar. (LMSW.)

A 10-foot-long full-scale mock-up of the A-12 nose section, including cockpit and canopy, was mounted on the 50-foot-tall pole to test various chine configurations. The nose section and forward fuselage of the airplane involved the most complex structure and was made from some of the thinnest materials. The chine assembly included radar-absorbent structures. (LMSW.)

Prior to test flights, the first A-12, known as Article 121, underwent fuel system and engine testing. In the above photograph, the airplane is undergoing an engine run adjacent to the North Taxiway. A fence (on the right) prevents observation from the main base area. The engine inlets are covered with screens to prevent ingestion of foreign objects, and fuel is fed from external tanks used only for ground tests. Below, fuel trucks and firefighters stand by. Upon fueling the airplane for the first time, 68 leaks developed as a result of problems with tank sealant. In order to stay on schedule, Lockheed crews had to strip the sealant and replace it with one that was known to have inferior qualities at high temperatures. Fortunately, this was sufficient for initial test flights. (LMSW.)

Crews prepare to hook up start-carts in preparation for starting the engines. Because the intended power plant, the Pratt & Whitney J58, was not yet available, the prototype was initially equipped with the less capable J75 and the moveable inlet spikes were fixed in place. The airplane is seen here on the north apron. Hangar Five is visible on the left and Hangar Six on the right. (LMSW.)

Article 121 rolls down the north taxiway toward the parking ramp, providing a good view of the aft quadrant. Control surfaces on the trailing edge are comprised of inner and outer elevons separated by the engine nacelles and twin canted, all-moving rudders. (LMSW.)

Crewmen assist Lockheed test pilot Louis Schalk with preflight preparations in Article 121. Kelly Johnson can be seen leaning into frame on the far right. The A-12 had three first flights. On April 25, 1962, the plane left the ground during a high-speed taxi run and flew for about a mile at about 20 feet altitude. The first real flight took place the next day and lasted 30 minutes. (LMSW.)

Schalk flew again on April 30 for the customer's official first flight. He was airborne for 59 minutes, attaining a top speed of 340 knots and a peak altitude of 30,000 feet. After conducting a stability and control check, Schalk declared the A-12 responded well in all flight regimes. (LMSW.)

Louis Schalk smiles at a comment from FAA chief Najeeb Halaby while shaking hands with former CIA Deputy Director for Plans Richard M. Bissell Jr., who had overseen the U-2 and the early stages of Project Oxcart. Other VIP observers seen here include Lockheed chairman Courtlandt Gross and NRO director Joseph Charyk (second from left). (LMSW.)

Article 121 vents fuel during an early test flight over Nevada. Until the J58 engines became available, Lou Schalk and fellow Lockheed test pilot Bill Park explored the airplane's low-speed performance envelope. (LMSW.)

Unlike most of the A-12 fleet, Article 121 never received anti-radar treatments. It retained all-metal chine and edge assemblies, described by Kelly Johnson as "titanium falsies," throughout its career as a dedicated test bed for airworthiness and handling qualities, envelope expansion, airframe/power plant integration, subsystems, and propulsion. (CIA.)

Ground crewmen prepare Article 121 for a test flight beside a Lockheed F-104 chase plane. Two McDonnell F-10B chase planes are visible in the background. Chase aircraft accompanied the A-12 only during low-speed phases of flight. Once it was equipped with J58 engines, the A-12 was capable of cruising at Mach 3 speeds and altitudes up to 90,000 feet. (LMSW.)

Between April 1962 and June 1968, Article 121 was flown 322 times, acquiring 418.2 flight hours. Initially equipped with two J75 engines, it was flown with one J75 and one J58 for the first time on October 5, 1962, and with two J58s on January 15, 1963. Lockheed test pilot Jim Eastham piloted the first sustained Mach 3 flight on February 3, 1963. In January 1964, Eastham achieved a speed of Mach 3.3 in Article 121 and cruised at the design Mach number of 3.2 for 15 minutes. The basic A-12 design eventually spawned a series of variants with characteristics optimized for specific missions. (LMSW.)

In early 1964, nearly the entire fleet of Mach 3 aircraft was lined up on the north apron for a family portrait. Article 121 is in the foreground, followed by Article 124 (the A-12T trainer) and Articles 125 through 130, as well as two YF-12A interceptors, Article 1001 and 1002. The EG&G radar building is visible in the background. (LMSW.)

Article 130 arrived at Area 51 in December 1963. Seen here in front of the Southend hangars, the aircraft displays an early paint scheme featuring bare metal finish with black paint applied to the nose, chines, inlet spikes, tails, and leading and trailing edges of the wings to more effectively radiate heat. Oxcart project pilots used this aircraft for training and proficiency flights. (LMSW.)

Several A-12s were lost during the course of the program. On May 24, 1963, project pilot Ken Collins was forced to eject during a subsonic engine test sortie in Article 123 after ice built up in the pitot tube, causing erroneous airspeed readings and loss of control. Collins bailed out after the jet stalled, pitched up, and entered an inverted flat spin. The secret jet crashed off-range, south of Wendover, near the Utah-Nevada border. After determining that Collins was safe, crews from Area 51 secured the scene and scrambled to recover the debris. Secrecy of the Oxcart program was maintained by telling the press that a Republic F-105 had crashed. (CIA.)

An A-12 stands by for departure clearance. In the background, an F-101B waits to take the runway while another A-12 is readied on the South Trim Pad. During the Oxcart program, the base population peaked at 1,835 with three work shifts per day. In May 1967, three A-12 aircraft were deployed to Kadena, Japan, for Operation Black Shield reconnaissance flights over Vietnam and North Korea. (LMSW via Jim Goodall.)

By late 1964, most of the airplanes were painted entirely black, and this paint scheme was adopted for all subsequent variants, earning the nickname "Blackbirds." Article 128, seen here on the runway, was lost due to fuel starvation on January 5, 1967, while returning from a routine training flight. Project pilot Walter Ray ejected but failed to separate from his seat and was killed. (LMSW.)

The Oxcart trainer, Article 124, arrived at the test site in November 1963. It had a second cockpit, for an instructor pilot, behind and above the student cockpit. Lockheed designers designated the trainer A-12T, but pilots' flight records (AF Form 5) list the airplane's Mission Design Series designator as TA-12, a more conventional nomenclature for contemporary training aircraft. In the above photograph, Article 124 is parked next to the Lockheed control tower at the northeast corner of Hangar Five. Below, the A-12T awaits takeoff clearance for a training flight. (RRI.)

Unlike the rest of the A-12 fleet, the A-12T was never painted overall flat black, but instead it was left primarily in natural metal finish. This airplane retained J75 engines throughout its service life and thus never attained Mach 3 speeds. The A-12T was nicknamed "the Titanium Goose." (LMSW.)

Even beneath an oxygen mask, Kelly Johnson appears to have a pleased expression as he sits in the student cockpit of the A-12T—this was Johnson's only flight in his creation. Thought to have taken place in December 1964, the exact date remains a mystery because, uncharacteristically, Johnson failed to record the event in his journal. (LMSW.)

In 1960, with the assistance of the CIA, the Air Force contracted with Lockheed to build three prototypes of an interceptor version of the A-12. Known as the AF-12, the design included a second crew position, launch capability for air-to-air missiles, and nose-mounted radar. The program, called Project Kedlock, was funded entirely by the Air Force. In 1962, the Department of Defense instituted a common designation system for military aircraft, under which the AF-12 became the YF-12A. Jim Eastham piloted the maiden flight of the first prototype, Article 1001, on August 7, 1963, while Lou Schalk flew chase in an F-104. (CIA/LMSW.)

The AF-12 nose configuration, designed to accommodate the radar, significantly altered the airplane's aerodynamics and directional stability. Engineers resolved the problem by adding two small ventral fins to the engine nacelles and a large, hydraulically powered folding ventral fin on the centerline of the aft fuselage. Because of its size, the fuselage fin had to be folded to one side prior to takeoff and landing. (LMSW.)

The YF-12A underwent extensive ground testing and preflight preparation before each test flight. Pilots found that the airplane performed as well as the A-12 and had similar handling qualities. It was capable of launching several Hughes GAR-9/AIM-47 missiles from a weapon bay in the forward fuselage while flying at Mach 3 speeds. (LMSW.)

Article 1001 and Article 1002 are photographed here on the Compass Rose pad on the south taxiway. President Lyndon B. Johnson publicly announced the existence of the YF-12A on February 29, 1964. At Kelly Johnson's request, however, he called the airplane A-11, because that designation denoted the non-anti-radar design. Within minutes of the president's announcement, the Articles 1001 and 1002 were on their way to Edwards Air Force Base. The third, Article 1003, completed its maiden flight 12 days later and went on to set several official speed and altitude records. (LMSW.)

The second YF-12A, Article 1002, was first flown on November 23, 1963. Only three of the Mach 3 interceptor prototypes were ever built. At the time, Secretary of Defense Robert McNamara was engaged in a bitter feud with the Air Force over appropriation of defense funds, and he specifically targeted planned production of the production model, the F-12B, in cost-cutting measures. (LMSW.)

Oxcart project pilots and Roadrunners administrative personnel pictured are, from left to right, Ronald "Jack" Layton, Dennis Sullivan, Mele Vojvodich Jr., Burton S. Barrett, Jack Weeks, Kenneth Collins, Walter Ray, Brig. Gen. Jack Ledford (CIA Headquarters), William Skliar, Cy Perkins, Col. Robert J. Holbury, John Kelly, and Col. Hugh "Slip" Slater. In July 1966, Slater succeeded Holbury as commander of the 1129th SAS and Area 51. (CIA.)

Lockheed developed a ramjet-powered, unmanned reconnaissance drone, code-named "Tagboard," to be launched from an A-12 variant. Articles 134 and 135 were built equipped with a rear seat for a launch systems operator and a dorsal launch pylon. Because Kelly Johnson saw the two vehicles as "mother" and "daughter," he designated the mother ship M-21, the drone D-21, and the mated combination MD-21. Above, fit checks of the mated MD-21 configuration are taking place using Article 134 and Article 501, the first production D-21. Below, test-flight mechanic Eugene "Red" McDaris works in the cockpit during preparations for a captive test flight. (RRI.)

Although the CIA had primary management responsibility for Tagboard, it was largely an Air Force program. The first M-21 was delivered to Area 51 for initial checkout work on August 12, 1964, and the maiden flight of the MD-21 took place on December 22. Each D-21 was to be used only once, following a programmed route and depositing a recoverable camera package before self-destructing. The drone was propelled by a Marquardt XRJ43-MA20S-4 ramjet engine that could operate continuously for more than 1.5 hours. Until this time, no ramjet had ever powered a vehicle for longer than 15 minutes. Frangible cones covering the inlet and exhaust were pyrotechnically jettisoned prior to engine start. (LMSW.)

Article 134 carries the D-21 above the snow-covered Nevada landscape. This airplane, used only for captive-carry test flights and as a chase aircraft for Article 135, made 94 flights before being placed in storage at Lockheed's Palmdale facility on September 29, 1966. (LMSW.)

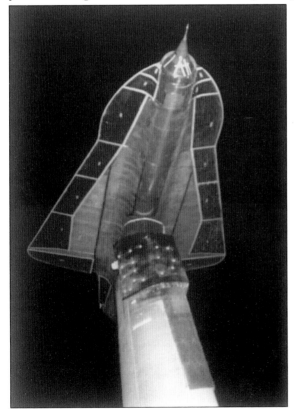

A full-scale D-21 mock-up saw service as a pole-mounted model for RCS measurements. The D-21 was about 43 feet long with a 20-foot wingspan. Because the design included the same type of anti-radar treatments as the A-12, the drone had a very small radar cross-section. According to James A. Cunningham Jr., deputy director of the CIA's Office of Special Activities, "It tracked like a pigeon." (LMSW.)

Captive tests were followed by separation trials. Article 135, the second M-21 built, successfully launched three drones. The airplane is seen here preparing for takeoff with Article 505 on June 16, 1966. Due to the hazardous nature of the tests, the drones were launched over the Pacific Ocean. During a July 30 launch attempt, a drone struck the tail of the M-21 just after separating from the launch pylon, resulting in the loss of both aircraft. Pilot Bill Park ejected safely and was rescued 150 miles off Point Mugu, California. Launch systems officer Ray Torick ejected but drowned before rescuers arrived. As a result of the tragedy, Kelly Johnson cancelled further use of the MD-21. (RRI.)

In the wake of an April 1966 order for a second batch of drones, Kelly Johnson proposed building a modified drone, eventually designated D-21B, that would be launched from beneath the wing of a B-52H bomber. This, he said, would provide greater safety, reduced costs, and expanded deployment range. In order to propel the D-21B to ramjet ignition speeds around Mach 3, the drone required a rocket booster for the initial flight phase following launch. After the MD-21 tragedy, all remaining D-21 drones were modified to the D-21B configuration, and two B-5H aircraft were configured as launch platforms under Project Senior Bowl. An unofficial first flight occurred when Article 501 was accidentally dropped due to a mechanical failure shortly after takeoff from Area 51 during a captive functional check flight on September 28, 1967. (LMSW.)

The first planned launch took place in November 1967, and test flights continued through July 1969. The B-52 typically carried two drones, each weighing approximately 11,000 pounds and equipped with a 44-foot-long Lockheed DZ-1 booster with an A-92 solid-fueled rocket motor. The booster weighed 13,286 pounds and produced 27,300 pounds of thrust during an 87-second burn. D-21B drones flew 12 test missions and four operational sorties. Each flight cost approximately $5.5 million, and the operational missions failed to produce expected results because the camera packages were not recovered. The program was cancelled in July 1971 and all remaining drones placed in storage. (LMSW.)

In 1968, Area 51 hosted the first of many evaluation programs involving foreign aircraft, mostly types flown by the Soviet Union and its allies. Managed by the Air Force's Foreign Technology Division, Project Have Doughnut was a joint Air Force/Navy technical and tactical evaluation of a captured Iraqi MiG-21F-13 on loan from the Israeli Defense Forces Air Force. To conceal the aircraft's identity in unclassified documents, the MiG-21 was referred to as a YF-110B. It was transported to Area 51 in a cargo plane on January 23, 1968, and unloaded inside Hangar Five. Assembly began the next day, and all systems were inspected, adjusted, repaired, and operationally checked. (USAF.)

Air Force Systems Command (AFSC) personnel evaluated performance and handling qualities of the MiG-21. Maj. Fred J. Cuthill, chief of the Air Force Flight Test Center's Fighter Branch, served as AFSC project officer. Several pilots from Naval Air Test and Evaluation Squadron 4 (VX-4) and the Air Force Tactical Air Command conducted tactical evaluations of the aircraft. Comdr. Thomas J. Cassidy Jr. served as project officer from VX-4 and Lt. Col. Joe B. Jordan as TAC project officer. Jordan made the first flight on February 8, 1968. Photographs of the airplane were taken from every angle for later use in briefings to operational combat pilots. (USAF.)

Maj. Gerald D. Larson, TAC, observes as Fred Cuthill prepares for a test flight. Several other pilots also evaluated the YF-110B. The results were compiled in several reports outlining aircraft performance, propulsion, system and subsystem characteristics, and design features, as well as measurements of infrared and radar signatures, engine modulation, and acoustic measurements. (USAF.)

The YF-110B awaits takeoff clearance on the runway at Area 51. The airplane proved to be rugged and easy to maintain. The small jet was highly maneuverable and difficult to spot in combat maneuvers because its engine did not produce a heavy smoke trail. (USAF.)

The YF-110B was evaluated against nearly all types of US combat aircraft in order to develop combat tactics for use against adversaries, particularly, at the time, those in Southeast Asia. The results of Project Have Doughnut helped turn the tide of battle in the air war over Vietnam. (USN.)

Looking east, with Whitesides Hill in the distance, the YF-110B banks over Groom Lake. By March 30, 1968, the airplane had completed 102 sorties totaling nearly 77 flight hours. It took five days to completely disassemble the aircraft, which was then shipped back to Israel. In 1972, this airplane was returned to Area 51 for additional testing, joining a growing fleet of foreign aircraft at Area 51. (USAF.)

In 1969, two MiG-17F aircraft captured by Israel from Syria were delivered to Area 51. The first, code-named Have Drill and designated YF-113A, was delivered disassembled to Groom Lake on January 27. Lt. Col. Wendell H. Shawler, chief of the 6512th Test Squadron's Special Projects Branch at Edwards Air Force Base, served as project manager. VX-4 test pilots Lt. Cmdr. Foster S. "Tooter" Teague and Lt. Cmdr. Ronald E. "Mugs" McKeown drew up a test plan. Fred Cuthill made the first functional check flight in the YF-113A on February 17, 1969. By the middle of May, a handful of pilots had completed 172 technical and tactical evaluation sorties in this airplane. (USAF.)

The second aircraft, code-named Have Ferry and designated YF-114C, was delivered to the test site on March 12, 1969, and had its first functional check flight on April 9. The two MiG-17s were flown against numerous US aircraft types during the tactical evaluation. In a little over a month, the test team flew 52 sorties in the YF-114C, including dual missions with the YF-113A. Selected instructors from the Navy's Top Gun school were chosen to fly against the MiGs for familiarization purposes, and during the 1969 Tailhook Convention in Las Vegas, several Navy admirals visited Groom Lake for orientation flights in the foreign aircraft. On another occasion, Marine Corps Gen. Marion Carl logged 1.7 hours of flight during two sorties. The YF-114C remained at Area 51 for use in later evaluation programs. (USAF.)

The Have Drill YF-113A is pictured above landing on the Area 51 runway following an evaluation flight. This was the only airplane equipped with test instrumentation for obtaining quantitative data, primarily in a clean configuration (meaning no external stores). Limited quantitative data were obtained on the airplane with two 106-gallon external fuel tanks. Below, the Have Ferry YF-114C leads a two-ship formation. Although the YF-114C served primarily as a backup for the YF-113A, a total of 25 dual tactical missions were flown. Following completion of the project, the YF-113A was returned to Israel, but the YF-114C remained for use in follow-on programs. (USAF.)

Capt. Charles E. Trapp Jr. delivered this H-43B helicopter to Area 51 in September 1962. It was used for mountain-top construction and maintenance projects, search and rescue, survival training, security, and other utility purposes. In December 1962, Trapp flew airborne alert when Pres. John F. Kennedy landed at Indian Springs in *Marine One* for a visit to the Nevada Test Site. (RRI.)

Crewmen attach a load to the H-43B for use in a construction project atop Bald Mountain or Papoose Mountain. Because there were no roads, the helicopter was the only way to get people and equipment to the remote sites. Fire Station No. 2 is visible in the background, and the Personal Equipment Building is at left. The tower is part of the parachute rigging and repair facility. (RRI.)

Trapp uses the H-43B to lift a steel Conex container for a construction project. The helicopter had many other uses, including recovery of 1,000-pound camera packages dropped in support of D-21 development and hovering at 13,000 feet with a spherical metal radar target hanging below the chopper on a 3,000-foot-long cable. Hangar Eight, in the background, was used for maintenance of Area 51 support aircraft. (RRI.)

The H-43B hovers over the Bald Mountain antenna site. Construction required hauling 30,000 pounds of wet cement, 1,000 pounds at a time, to the 9,380-foot peak. The helicopter was also used to transport metal poles, welding equipment, generators, and tools, as well as food and water for the construction crew. (RRI.)

Firefighters train at the Crash Pit using jet fuel and a metal structure to simulate a downed airplane. Captain Trapp brought the H-43B in low, using the rotor wash to cool the area, clear away smoke, flatten the flames, and spread retardant foam to enable firefighters to extinguish the blaze using water cannons and hand lines. (RRI.)

A UH-1F was used during the final stages of construction at the Papoose Mountain site. Charlie Trapp and M.Sgt. Bill "Flash" Walters picked it up from the Bell Helicopter plant in Hurst, Texas, and flew it to Area 51. The UH-1F was faster than the H-43B and had doors on both sides of the cabin, which made loading and unloading passengers and cargo easier and safer. (RRI.)

Workers secure the Papoose Mountain antenna while it is held steady beneath the UH-1F. This helicopter was also used for search-and-rescue missions, including one for Article 125, the A-12 that crashed in January 1967. Trapp and his crew used the UH-1F to retrieve the pilot's remains and various parts of the aircraft. (RRI.)

Area 51 had a fleet of six F-101B and two F-101F Voodoos to provide chase and support during Project Oxcart. On September 26, 1967, Lt. Col. James S. Simon Jr. crashed an F-101B on approach to the Area 51 runway while flying chase during a night sortie of the A-12T. (RRI.)

Built as a standard F-104A in 1958, this airplane was converted to the F-104A/G configuration and used for B61 special weapons shape drop tests at Tonopah Test Range (Area 52). In July 1962, it was returned to Lockheed for use as a chase aircraft for the A-12, MD-21, and YF-12A at Area 51. (LMSW.)

By the beginning of the 21st century, the Groom Lake facility had expanded considerably. Many of the old buildings and trailers had been removed or replaced. Use of the Oxcart runway, which had been extended at the southern end by 4,600 feet, was eventually discontinued. A new concrete airstrip was built in 1991, and in 2001 a taxiway near the Southend was converted into a 5,000-foot-long airstrip. Crosswind runways on the lake bed were maintained, but one was eventually moved a mile farther south. Storage for base water and fuel supplies and other infrastructure was upgraded. New dormitories replaced Babbitt Housing, and several new hangars were built. The base population grew to include approximately 500 military and civilian DoD personnel and nearly 2,000 contractors. (Landsat.)

Three

DREAMLAND
A NATIONAL TEST FACILITY

In August 1961, the Federal Aviation Administration established restricted airspace above the Nevada Test Site and Area 51 that prohibited unauthorized overflights below 60,000 feet. This effectively shielded the Groom Lake operation from observation by all platforms except satellites. Aircraft approaching Groom Lake first entered the Yuletide Special Operations Area under control of Area 51 personnel located at the Nevada Test Site command post at Yucca Flat. Once cleared, they were then were handled by Jupiter or Boxer Control at Area 51.

During the late 1960s, Yuletide SOA was renamed Dreamland. After further restructuring of the airspace, this became the radio call sign for clearance into a 440-square-mile box surrounding Groom Lake at the heart of the Switzerland-sized Nevada Test and Training Range.

In 1977, the CIA transferred responsibility for Area 51 to the Air Force Flight Test Center. Though the agency no longer required use of the base, Red Hats activity alone accounted for 25 percent of all AFFTC sorties, and plans were underway for testing a variety of new aircraft equipped with revolutionary anti-radar capabilities known as "stealth." Under Project Score Event, improvements costing nearly $200 million were made at the secret base. An Air Force site manager and staff initially oversaw operations, but on April 1, 1979, the AFFTC commander formally designated and activated the unit as the 6516th Test Squadron under the supervision of the 6510th Test Wing. To shorten the chain of command, this order was revoked on May 3, and the unit instead became Detachment 3, AFFTC. Over the next 10 years, the detachment grew into a wing-sized organization that oversaw a national test facility responsible for conducting numerous classified test programs.

The name "Area 51" gradually fell into disuse, replaced simply by "Det 3." In some documents, the base has been called the National Classified Test Facility. By any name, the Groom Lake test site remains a valuable asset in development of aerospace vehicles and weapon systems. There, workers toil in relative isolation and obscurity to prove revolutionary technologies, enhance the readiness of the nation's war fighters, and support national defense requirements.

Testing of foreign aircraft continued and expanded throughout the 1970s and beyond, with integrated involvement of both Air Force and Navy personnel. The test organization initially consisted of one pilot and six maintenance personnel, limiting their activities to part-time or additional-duty efforts. By 1977, the project had grown to the point that the Air Force formed two squadrons, known as the Red Hats and Red Eagles, to perform technical and tactical evaluations of a variety of aircraft types. Program security requirements spawned a slew of oddball designations for test aircraft, including YF-110C, YF-110L, YF-112C, YF-113B, YF-113E, YF-114D, and YF-116A, and many more. Testing of these "classified prototypes," as they were euphemistically dubbed, became the Groom Lake test site's highest-priority program. (USAF via Steve Davies.)

Technical evaluations included experimental and developmental testing to explore performance characteristics and to evaluate handling qualities, propulsion, and avionics, as well as detailed examination of design features, airframe structure, subsystems, and equipment. Tactical exploitation included comparative evaluation of US and foreign types in simulated air-to-air combat to determine the effectiveness of existing tactics and to develop new ones. The results were briefed to National Command Authorities, Congress, chiefs of staff, commanders, and combat crew members in order to provide insight into the capabilities of potential adversaries. In the most extensive application of lessons learned, the Red Eagles, based at Tonopah Test Range, exposed nearly 6,000 US fighter pilots to simulated combat missions against MiG-17, MiG-21, and MiG-23 variants under Project Constant Peg. (USAF via Steve Davies.)

Lockheed's Have Blue experimental survivable test bed arrived at Area 51 in November 1977. The tiny demonstrator was the first airplane designed to be virtually invisible to radar—a technique officially termed "low observable" but commonly referred to as "stealth." The craft's boxy shape resulted in an exceptionally low radar cross-section. The first vehicle, seen in the photograph above, was flown to demonstrate basic handling characteristics. Officials monitored the first flight from the White House Situation Room and Tactical Air Command Headquarters at Langley Air Force Base, Virginia. A second Have Blue demonstrator, below, served as a low-observable test bed. Whenever either aircraft was exposed outside of its hangar, non-cleared personnel at the base were sequestered to prevent them from seeing it. (LMSW.)

In a humorous moment, Lockheed test pilot Bill Park demonstrates his solution to Have Blue's braking problems. Injuries sustained while ejecting from Blue-01 in May 1978 ended Park's flying career. Air Force test pilot Ken Dyson completed the flight program in Blue-02. During a sortie in July 1979, Dyson was also forced to eject. Although both vehicles were lost, sufficient data had been collected to justify a follow-on program. (LMSW.)

In the late 1970s, Area 51 looked much as it had during Project Oxcart. Since the Red Hats operation occupied the North Ramp hangars, Lockheed test teams housed their projects at the Southend, which came to be known as "Baja Groom Lake." Test flights usually took place during the first hour after sunrise before most non-permanent test-site workers had arrived. (Author's collection.)

In 1972, EG&G Special Projects was awarded a contract to provide commuter air service to Area 51. Based at McCarran Airport in Las Vegas, the fleet initially consisted of two DC-6B passenger transports that remained in service until 1981. This photograph of N6583C was taken at McCarran International Airport on May 16, 1973. (Peter Nicholson.)

Due to increasing levels of work activity at Area 51, EG&G purchased three Boeing 737-200 widebody jets from Western Airlines in 1980. Seen here in March 1981, N4508W taxis at McCarran. Using the call sign "Janet," the EG&G fleet eventually grew to include six 737s. The generic paint scheme, white with a red stripe, drew unwanted attention from airliner fans marveling at the concept of a "secret airline." (Michael Haywood.)

In 1987, two 19-seat Beechcraft B1900C turboprops (above) joined the commuter fleet, followed two years later by three Beechcraft 200C Super King Air 13-seat aircraft. Referring to their differing paint schemes, workers nicknamed the 737s "the red and whites" and the Beechcraft models "the blue and whites." In March 2004, a B1900C en route to Tonopah Test Range from the National Classified Test Facility at Groom Lake crashed when the pilot suffered a heart attack. All four passengers aboard also perished. Subsequently, flight rules were modified to require two pilots onboard. The aircraft are registered to the Department of the Air Force. (DLR.)

In October 1978, Lockheed conducted the first test of a stealth cruise missile, code-named Senior Prom. Six prototypes were built, each somewhat resembling a subscale, unmanned version of the Have Blue. These demonstrator models were launched and controlled from NC-130B aircraft. Senior Prom test articles and launch aircraft were housed in Hangar 17 at Area 51, where technicians experimented with several different configurations. The earliest model (pictured above) had no conventional tail surfaces, only winglets and a rectangular ventral fin that would have been omitted from the production model. Later versions had a V-tail configuration and narrow wings of constant chord. Although the demonstrators had fixed wings, a production model would have had folding wings for facilitating internal carriage. (LMSW via Jim Goodall.)

Originally, the Senior Prom test plan called for each aircraft to be used only once, but Lockheed engineers equipped the vehicles with a recovery system using a ballistic parachute and inflatable ventral landing bag. The six craft ultimately made as many as 14 flights, with one air vehicle reportedly flying nine times. For RCS testing, the Senior Prom aircraft was coated with radar-absorbent material and flown over the EG&G test range. According to Bill Fox of Lockheed, the stealthy drone flew over the base at an altitude of about 500 feet without generating a significant radar return. There were few witnesses to the event because personnel not cleared for the sight-sensitive program were sequestered in the chow hall. Despite its success, the program was terminated in 1981. (LMSW via Jim Goodall.)

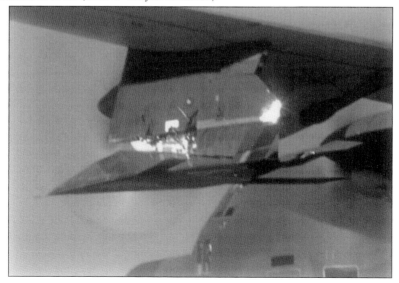

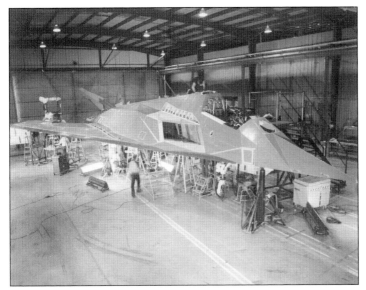

Lockheed's Senior Trend was developed using data from Have Blue. Following delivery to Groom Lake in January 1981, the prototype, Article 780, was reassembled and checked out in preparation for flight tests. A total of five full-scale development aircraft, designated YF-117A, were produced along with 59 production F-117A airframes. (LMSW.)

A firefighter observes as Lockheed technicians conduct the first engine run for Article 780 inside the hangar on April 28, 1981. The YF-117A resembled a faceted arrowhead. The airplane's conventional aluminum and titanium fuselage was equipped with radar-absorbent edge structures and coated with radar-absorbent material. The engine inlets were screened to reduce radar cross-section. (LMSW.)

It took five months to reassemble the YF-117A and ready it for its maiden flight. Because the airplane was considered sight sensitive, camouflage netting was hung over the open hangar door to prevent non-cleared personnel from seeing the stealthy prototype. After finding a scorpion in one of the Southend hangars, the Senior Trend Joint Test Force (JTF) took the nickname "Baja Scorpions." (LMSW.)

Seen from the roof of Hangar Nine, the YF-117A taxis out for its maiden flight. The airplane was crudely painted in a desert camouflage scheme, an apparent attempt to hide its shape from unauthorized observers. The flat exhaust ducts were placed in such a way as to reduce the aircraft's infrared signature (LMSW.)

Article 780 first flew on June 18, 1981, with Lockheed test pilot Harold "Hal" Farley at the controls. The planned 30-minute sortie was cut short after just 13 minutes due to a canopy warning light and overheating in the exhaust duct. In a tradition started by the Red Hats, all test flights at Area 51 used the call sign "Bandit" followed by individual numbers for each pilot. Farley initially used the call sign Bandit 01 but later adopted Bandit 117 for the remainder of the Senior Trend project. (LMSW.)

Bill Fox, Lockheed site manager and engineering test manager for Senior Trend, douses Hal Farley with a bucket of water. This has long been a tradition for celebrating completion of a first flight. The JTF consisted of personnel from Lockheed (Advanced Development Projects–Skunk Works), Air Force Systems Command (AFFTC and the System Program Office), and Tactical Air Command. (Bill Fox collection.)

The Skunk Works crew celebrated the first flight of the YF-117A with an all-night party in the hangar. Ben Rich, Lockheed's vice president for advanced projects and Kelly Johnson's successor, is third from left in the second row. (LMSW.)

Article 780 was painted overall light gray to reduce its daytime visual signature. Ben Rich intended to deliver the production F-117A in gray, but Gen. William Creech, chief of Tactical Air Command, wanted the airplanes painted black. "You don't ask the commander of TAC why he wants to do something," Rich recalled, "He pays the bills. If the general had wanted pink, we'd have painted them pink." (LMSW.)

Senior Trend test pilots pose with the third YF-117A, Article 782, in front of the Southend hangars. From left to right are Hal Farley, Skip Anderson, Dave Ferguson, Tom Morgenfeld, Roger Moseley, Tom Abel, Jon Beesley, Paul Tackabury, Pete Barnes, Denny Mangum, and Dale Irving; other pilots included Bob Riedenauer and Skip Holm. (LMSW.)

Article 782 was painted in a patriotic flag motif for a change-of-command ceremony. On December 14, 1983, Jon Beesley made a low pass over the base in the YF-117A following an event in which Roger Moseley transferred command of the Senior Trend JTF to Paul Tackabury. The flag remained on the aircraft until March 1984. (LMSW.)

Rank has its privileges. Col. Ralph H. Graham, commander of Detachment 3, AFFTC, exits the cockpit of Article 787 on October 14, 1983, following his first orientation flight in the F-117A. Article 787 was used for initial operational test and evaluation and was nicknamed "Pete's Dragon." (USAF via Red McDaris.)

In the stressful world of black projects, workers need a place to unwind. Several options were available at Groom Lake, including Sam's Place (the main recreation facility, named for former Area 51 commander Sam Mitchell of the CIA) and the Conehead Bar in House 79, one of the old dormitories. The F-117A avionics engineers had adopted the name "Coneheads." Their patch included the hexadecimal numeral "4F" that, when converted to its decimal equivalent, equaled 79. Conehead bartenders, seen above, were always ready to offer a smile and a cold drink. Below, a worker throws darts. Note that the clock on the wall is upside down. (SFA.)

Numerous recreational activities were available to workers who remained at the remote location throughout the week, including outdoor sports. Continuing a tradition started by the 8-Ballers during Project Oxcart, softball remained a popular pastime at the secret base. The Ballpark Franks won the 1987 championship. (SFA.)

What hangar party would be complete without a live band? Workers at Groom Lake pooled their resources and talent to provide music at celebrations for various milestones. Despite stories of wild parties and drunken revelry, the professionalism with which test team members performed their jobs has never been questioned. (SFA.)

Northrop developed a unique demonstrator called Tacit Blue to test technologies for use in a stealthy battlefield surveillance aircraft. Although plans to put such a craft into production were abandoned, Tacit Blue provided data that aided in the development of several other weapons systems, including the B-2 advanced-technology bomber, AGM-137 Tri-Service Standoff Attack Missile, and E-8 Joint-STARS aircraft. (USAF.)

Northrop test pilot Richard G. Thomas made the first flight of Tacit Blue on February 5, 1982. The plane's unusual appearance earned it the nickname the "Whale." Mementos carried on the maiden flight included buttons, extra name tags, and commemorative postal covers (with a picture of a Metro train since the airplane's shape was classified). The name "Jack" on the button refers to company founder Jack Northrop. (Author's collection.)

Tacit Blue, also known as the YF-117D, made a total of 135 sorties in three years. Dick Thomas, the only contractor pilot, made 70 flights, including the 100th sortie on April 27, 1984. Ken Dyson was the first of several Air Force pilots to fly the unusual craft; others included Russ Easter, Don Cornell, and Dan Vanderhorst. (Northrop via Dick Thomas.)

Tacit Blue was the first stealth aircraft to feature curved surfaces for RCS reduction. Northrop built only one complete airframe and a partially completed shell that was available as a backup. The "whalers" housed the airplane in Hangar Eight. Dan Vanderhorst piloted the final flight in February 1985. (Northrop via Dick Thomas.)

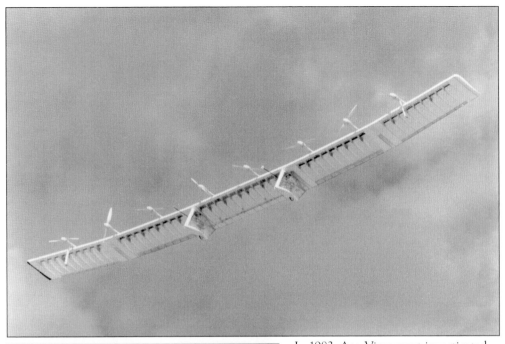

In 1983, AeroVironment investigated a concept for a high-altitude, solar-powered unmanned aircraft called HALSOL. In June and July, nine test flights were conducted using radio control and battery power, since the craft had not yet been equipped with autonomous flight controls or solar cells. These tests validated the vehicle's aerodynamics, but photovoltaic cell and energy storage technology were not sufficiently mature to make the idea practical at the time. (AeroVironment.)

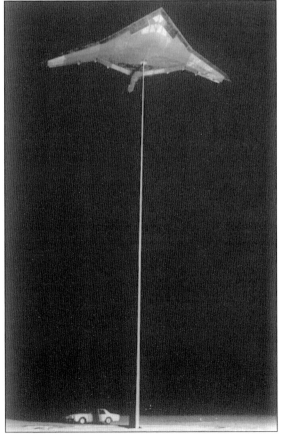

The EG&G radar range continued to provide vital support for numerous programs. Here, the "swimming pool" site sports a sophisticated pylon holding a scale model of Senior Peg, Lockheed's entry in the Advanced Technology Bomber competition. The Lockheed concept was rejected in favor of Northrop's Senior Ice design, which was later developed into the B-2. (LMSW.)

The Groom Lake radar complex grew to include a wide variety of reverse-engineered and acquired foreign electronic systems from a simple array of four Yagi antennas simulating a low-frequency search radar, above, to an Oborona (Tall King-C) identical to the one pictured below. These systems were given whimsical names such as Mary, Kay, Susan, and Kathy and were arranged to simulate the characteristics of a Soviet-style integrated air-defense system. An airborne RCS range known as Project 100, or the Dynamic Coherent Measurement System (DYCOMS), has been used to verify the radar signatures of stealth aircraft, including the F-117A, the B-2, and the F-22. (RRI/DoD.)

Most non-permanent base residents commuted to the test site on Monday and often stayed until Thursday or Friday. The base population saw a marked increase in support of the Senior Trend program, and it soon became obvious that the hooches erected during Project Oxcart were no longer satisfactory. In 1986, the old Babbitt houses (above) were replaced with brand-new dormitories. Residents had access to recreational facilities and cable television. (SFA.)

The Air Force supplied T-38A and AT-38B aircraft to support the various classified programs. The supersonic jets provided safety chase, as well as photographic and video support. Since most test articles were either single-seat or unmanned, the chase planes often carried test-flight engineers to monitor tests and provide real-time data analysis. (SFA.)

By November 1986, the EG&G fleet included six 737-200 passenger transports. This rare image shows all six parked on the Area 51 ramp adjacent to the passenger terminal, flight-crew lounge, and a security office where incoming passengers underwent a badge check. From the lower left, the aircraft appear to be lined up as N4510W, N4508W, N4515W, N7380F, N4529W, and N7383F. (DLR.)

The YF-118G Bird of Prey was a one-of-a-kind demonstrator designed and built by the McDonnell Douglas Phantom Works advanced research-and-development organization in St. Louis using company funds. The Air Force provided flight-test facilities, chase aircraft, engineering personnel, and one test pilot for the flight-evaluation program. After McDonnell Douglas merged with Boeing in 1997, the Boeing Company continued funding the project, which spanned eight years and cost $67 million. Only three pilots flew the Bird of Prey: Rudy Haug (McDonnell Douglas/Boeing), Lt. Col. Doug Benjamin (USAF), and Joseph W. Felock III (Boeing). Haug made the first flight on September 11, 1996. By the time the flight program ended in April 1999, a total of 38 missions had been flown—roughly one sortie per month. (USAF.)

Doug Benjamin was the only Air Force pilot to fly the YF-118G. Designers saw the Bird of Prey as a tool for providing state-of-the-art technology demonstration at the lowest possible cost. The airplane incorporated many off-the-shelf components as well as innovative concepts to reduce radar, infrared, and visual signatures. (USAF.)

Silhouetted against an early morning sky, the unusual shape of the Bird of Prey suggests something otherworldly. Although it looked like a futuristic fighter plane, the YF-118G had a thrust-to-weight ratio more like that of a cargo transport. The airplane's existence was declassified in 2002 because technologies and capabilities developed during the program had become industry standards. (USAF.)

The Ghost Squadron operates helicopters for mission support, search and rescue, and security at the National Classified Test Facility. In the 1980s, the squadron was equipped with three CH-3E Jolly Green Giant helicopters. A decade later, these were replaced with the MH-60G Pave Hawk. The choppers are most often seen performing perimeter sweeps to search for unauthorized observers. This problem became particularly acute in the mid-1990s, when rumors of UFOs turned the test-range boundary into a tourist mecca. Unusual sightings can undoubtedly be attributed to test and training activities. (Author's collection/Chuck Clark.)

The commuter fleet has undergone a few upgrades over the years. In the early 1990s, several of the older 737 models were replaced with converted Air Force CT-43A trainers outfitted to carry passengers. Like their predecessors, these airplanes were equipped with noisy turbojet engines. In 2008, the fleet began a nine-month transition in which all 737-200 models were replaced with turbofan-powered 737-600s that were registered to the Department of the Air Force. On weekends, all six are usually parked at a private terminal at McCarran International Airport. (DLR/Jakub Vanek.)

DISCOVER THOUSANDS OF LOCAL HISTORY BOOKS
FEATURING MILLIONS OF VINTAGE IMAGES

Arcadia Publishing, the leading local history publisher in the United States, is committed to making history accessible and meaningful through publishing books that celebrate and preserve the heritage of America's people and places.

Find more books like this at
www.arcadiapublishing.com

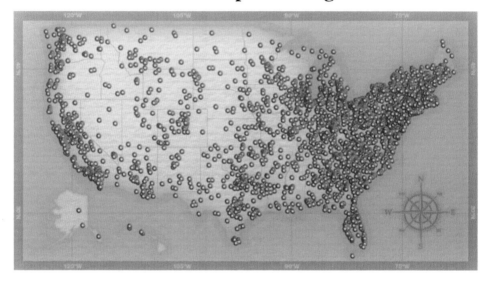

Search for your hometown history, your old stomping grounds, and even your favorite sports team.

Consistent with our mission to preserve history on a local level, this book was printed in South Carolina on American-made paper and manufactured entirely in the United States. Products carrying the accredited Forest Stewardship Council (FSC) label are printed on 100 percent FSC-certified paper.

MADE IN THE  **USA**